Paper-roll

Chromatography/Electrophoresis
(Large Scale Separation Technique)

Akintunde M Lawal

i

This is a non-fiction work. The author has represented and warranted full ownership and/or legal right to publish all the materials in this book.

Paper-roll Chromatography/Electrophoresis (Large Scale Separation Technique) by Akintunde M Lawal

Forward

This book is for the reading pleasure of scientists and the general public. Home remedy buffs who find it difficult to cope with the scientific terms used in this book should endeavor to read the glossary at the end of this book or any handy relevant dictionary/encyclopedia available to them for clearer understanding. Paper-roll chromatography and electrophoresis offer new techniques for separating, isolating or purifying large quantities of organic mixtures. Scientists wishing to collaborate with the author for further development of this technique should contact the author.

Dedication

To scientists, chemists, pharmacists, home remedy buffs, do-it-yourself apothecaries and laity whose quest to separate large quantities of pigments and natural products is insatiable.

CONTENTS

Introduction

This ebook is written for the purpose of sharing information on the theoretical and practical basis of another type of paper chromatography/electrophoresis technique with scientists and the general public. The author hopes that readers would appreciate the significance of this technique in the quest to evolve a cost effective method for the separation of large quantities of pigments or natural products.

Many potent organic compounds are found in nature but their use in the treatment of diseases have been curtailed by the expensive separation, isolation and purification techniques that are required for extraction and isolation of active components from biodiversity or synthetic mixture.

Paper-roll chromatography offers a more cost effective technique that is ideal for the separation/isolation of large quantities of pigments and other natural mixtures. However, before any description of this new technique is done, there is need to offer brief information on conventional separation techniques which are already available to scientists and home remedy enthusiasts.

Chromatography has been the scientific term used to describe all types of physical laboratory procedures that are used to separate mixtures of organic compounds. This technique is based on the partitioning of components of mixtures between mobile and stationary phases depending on the partition coefficient of each component. Ever since the technique was developed in 1900, there has been renewed interest at developing cost effective procedures for separating large quantities of organic mixtures.

The most common types of chromatography are paper and thin layer chromatography where small quantities of organic mixtures are separated or isolated.

Another type of separation technique, column chromatography is used for the isolation and purification of large quantities of organic mixtures on small or large scale basis. It allows the individual collection and separation of components of organic mixtures by running different suitable solvents through a packed column of alumina, cellulose or silica gel which acts as the stationary phase. The interaction between mobile and stationary phases follows the same principle as that of paper chromatography, hence the possibility of designing a hybrid technique as seen in paper-roll chromatography.

In the following pages an attempt is made to provide a brief description of conventional chromatographic techniques before the principle and procedure of paper-roll chromatography is given.

Types of Chromatography and Electrophoresis

There are different types of chromatography depending on the types of surface used, physical state of mobile phase, exchange mechanism and partition or adsorption mechanism. For the purpose of this book we shall restrict ourselves to paper, thin layer and column chromatography because Paper roll chromatography combines some attributes of all these techniques. This process can also be designed for electrophoresis.
In conventional paper chromatography, a spot or streak of the sample mixture is placed onto a strip of paper (stationary phase) which is placed in a sealed container with a shallow layer of suitable solvent (mobile phase). Polar components of the sample mixture lag behind while non-polar components move ahead with the solvent.

Electrophoresis is another technique for separating organic mixtures because of their ionic nature. Organic substance such as amino acids and small protein molecules exists as ionic molecules which can be separated in mixtures that are placed in an electric field. Most often, these molecules move through buffer solutions in gel or filter paper.

Types of Paper Chromatography

Conventional types of paper chromatography include,
Planar Paper chromatography
Paper Column chromatography
Circular paper chromatography

Description of these aforementioned techniques can be sourced from various books and journals. This book addresses the need to develop a new technique aptly termed "Paper-roll chromatography" that would allow the separation, isolation and purification of gram to kilogram amounts of organic mixtures for laboratory, health or industrial uses/applications.

Principles of Paper-roll techniques

Paper roll chromatography involves the concept of partition coefficient of organic components of mixtures between two immiscible solvents. Components adsorbed on a solid support matrix such as paper, silica and alumina are partitioned between minute molecules of solvents on the stationary phase and solvents in the mobile phase. Substances used as the stationary phase contain many free hydroxyl groups (OH) which make them polar. Interaction between molecules of organic mixtures to be separated and the stationary phase occur by hydrogen bonding between hydroxyl groups, water bonded cellulose paper (stationary phase) and hydrogen ions (mixture) retard the movement of components as the solvent moves up the paper by capillary action.

Paper-roll is made by wrapping the paper into a cylinder, stapling it at the top and bottom to hold the edge of the paper-fold in place. This roll can be designed in two ways either for the purpose of chromatography or electrophoresis as,

A roll of lengthy paper strip, or

A roll comprising of two lengthy paper strips sandwiching alumina or silica gel (for use in chromatography). Buffer solution alone or in combination with gel (for use in electrophoresis)

is sandwiched instead of silica gel.

When the lengthy paper strips are wrapped, the remaining one centimeter of the folded is secured onto the body of the roll by means of a staple pin, drawing pin or any suitable metal fastener.

In ascending order paper roll chromatography a lid is provided for the chromatography chamber or cylinder to achieve solvent vapor saturation which increases efficiency of the process. This technique is a hybrid between conventional paper, thin layer as well as column chromatography techniques. Hence is can be developed in ascending or descending order formats.

After developing the chromatogram, each separated band is formed at the position of the retardation or retention factor, Rf of each component of the organic mixture. This Rf is a ratio of the distance the component moves to the distance the solvent or mobile phase travels. Each component can be obtained by cutting the paper roll into segments along the blank areas between the various bands of components formed on the paper.

However, paper-roll electrophoresis involves using high direct current voltage to produce an electric field that separate organic compounds that are streaked on a paper roll that has been moistened with buffer solution. Conventional chromatography or electrophoresis is done with small amounts of organic materials for the purpose of determining comparative proportions of analytes in a sample mixture. However, Paper-roll chromatography or electrophoresis allows the separation of grams to kilograms amount of analytes without the attendant need for large amount of solvents and expensive stationary phase materials. Ascending order Paper-roll Chromatography: Ascending order paper-roll chromatography resembles paper strip, thin layer and column chromatography. A small amount of suitable solvent is poured into the chromatography chamber to form a shallow layer that is few millimeters deep. The lid is put in place while the chamber is allowed to stand for a lengthy period for optimal solvent vapor to saturate within the enclosure. This procedure allows for better development of chromatograms. To prepare alternating or sandwich layers of paper with silica gel, alumina, cellulose or calcium carbonate granules, a slurry of these adsorbent materials are made with little amount of water and spread as a thin layer of few millimeters thick on long strip of paper. A pencil line is drawn at 0.5-1.0 cm from the bottom of the chromatography paper or sandwiched paper. Then, a streak of sample mixture is applied on the paper along the line of origin, allowed to dry before the paper is wrapped around a suitable cardboard or polymer core. A pin (staple or office pin) is used to fasten the edge of the rolled paper onto the cylindrical body to prevent unfolding of the paper roll. The roll produced is dried in a large oven for a period ranging between 30 minutes to 24hrs at 40 to 100 degree Celsius (depending on length of paper strip/quantity of sandwiched adsorbent material) and subsequently hanged within the chromatography chamber containing any suitable solvent with a T-shaped glass, plastic or any suitable polymer handle. It should be ensured that the streak of sample on the paper roll is slightly above solvent level. The chromatography process is terminated and the paper roll removed from the chamber as soon as mobile solvent is near the upper edge of the paper roll. Separated components are identified by comparing their Rf values with that of a conventional strip paper chromatography that must have been done on the same sample in a previous procedure. Alternatively a dry paper strip of the same height but of shorter width could be pressed on the developed paper-roll when it is still fresh and wet with solvent. When separated and dried at room temperature, various Rf of separated components can be identified. A comparative side by side study of this paper is done with the paper roll to locate the band of each component. After retrieving the paper roll from the chromatography chamber and dried at room temperature, a sharp instrument such as blade or knife is used to cut the roll into segments along the lines between each separated band of components. Bands of components are identified by fluorescence in ultraviolet light or any other suitable indicator. Suitable solvents are now used to dissolve and extract each component from the segments produced. Descending order Paper-roll Chromatography: In descending order paper-roll chromatography, the roll is within a tube comprising of two half cylinder tubes that are clasped around the roll with the aid of two clamps...at the upper and lower ends, a retort stand holds the paper roll assembly in vertical position. This tube is designed to taper into a narrow opening at the lower end. The sample mixture is dissolved a small amount of the mobile phase and pipette onto the upper edge of the paper roll. More quantities of the mobile solvent are now poured into the roll at the upper part and allowed to move down the roll. The process is stopped once the mobile phase has reached the lower edge of the roll. Alternatively, more mobile solvents could be added at the top to allow elutates of components to come out of the lower end, in this case a tap is desired at the lower part of the detachable cylinder. This method may include a Flash paper roll chromatography wherein positive pressure is applied to drive the motion of the mobile phase by the use of a pump. The amount of solvent required is reduced and a faster separation technique can be achieved. Paper-roll, detachable glass tube assembly is clamped vertically by using a retort stand. This glass or plastic tube must be tapered at the lower end. It should be noted that the paper roll should not be torn or jagged in order not to allow air to be trapped within the paper strands as to avoid any disruption in the bands of components formed on the chromatogram. A small amount of sample is dissolved in five to ten drops of a polar solvent before being dropped with a pipette onto the upper edge of the paper roll. A sandwich paper-roll comprising of two paper strips between which a thin layer of silica gel or alumina has been laid and rolled into a cylinder can also be used to form a thicker roll that can separate even larger quantities of sample mixtures. Solvents used for paper-roll chromatography should be less polar than compounds that are being separated and solvents are used in order of their polarity from non-polar to polar. A single solvent can be used to run the chromatogram until the mobile phase is nearer to the lower edge of the paper roll. The process is stopped, detachable glass tube dismantled and the paper roll removed for drying and marking of the bands of components which are identified by using ultraviolet bands or stains on a paper strip that has been pressed on one side of the paper roll when it was still wet with solvent. This strip is laid side by side with the chromatogram in order to identify the movement of bands and the area in which the paper roll will be cut into segments. In more complex separation techniques, series of pure solvents or mixtures of solvents from non-polar to polar can be used to run the paper-roll or sandwich paper roll to elute different components out. Eluted compounds can be detected by using peculiar color of separated components, visible spectrometry, gas chromatography or thin layer chromatography. Each eluted fraction is evaporated in a rotary extraction to distil off the solvent and recovered the pure component. Paper roll can be varied in length a diameter in order to select the appropriate dimension required for the separation of any particular type of organic mixture.

Paper-roll Electrophoresis:
Paper-roll electrophoresis is useful for the separation of small molecules that are charged and exist as ionic organic compounds, e.g. amino acids and proteins. In this technique a lengthy paper strip is streaked with a sample of the mixture to be separated, moistened with buffer and rolled up with alternating strips of thick carbon paper electrodes at the edges. This paper roll is then immersed into a buffer reservoir and a supply of high voltage direct current of 150V, 1ampere or 350V, 35milliampere is supplied to the carbon paper terminals.

Bands of separated amino acids and proteins with small molecules migrate. The process is stopped after some time and the movements of components are determined by a comparison with a previous similar paper electrophoresis on a strip of paper. The paper roll is cut into segments and suitable solvents are used to extract each component for the cut out segments. This process may be accomplished with the moistened paper-roll that is either simple or sandwiched with gel or cellulose and oriented in the horizontal or vertical plane.

Uses and application

Paper-roll Chromatography:

To determine the number of components in an organic mixture.

To purify organic substances.

To obtain large quantities of components of organic mixtures.

As a preparative technique in forensic and toxicological analysis.

For analysis of many organic substances.

As a hybrid technique that combines the attributes of paper, thin layer and column chromatography.

It produces faster mobile phase runs, relatively wider separations, and allows choice of different adsorbents. Between alternating paper folding or sandwiched adsorbent paper rolls. It also permits better resolution and quantification of components in sample.

Paper-roll electrophoresis:

To determine number of amino acids and small protein molecules in a mixture.

To purify gram quantities of amino acids and proteins.

As a preparative technique in forensic medicine and toxicology.

As a preparative technique in the scientific analysis of natural compounds.

Future developments

Paper-roll Chromatography:

The following procedures require further development in terms of designing specific equipment to perform easier or automated tasks in laboratory or industrial settings:
Streaking samples on unrolled, lengthy paper strip and rolling up streaked paper strip onto cardboard or thin polymer core.
Fastening end of paper roll to prevent unfolding or paper and hooking paper roll onto T shaped chromatogram handle.

Placing paper roll-T handle into large chromatogram vat.

Pouring suitable solvent into large vat.

Closing lid of large vat after paper roll has been mounted in.

Removing developed paper roll chromatogram. Identifying bands of separated components.

Cutting paper roll chromatogram into segments.

Eluting separated segmented components into suitable solvent.

Designing an automated system that can handle all the procedures required to separate any mixture by using paper roll chromatography.

Designing paper-roll sandwiched with silica gel, alumina or calcium carbonate for separating large quantities of proteins.

Paper-roll Electrophoresis: Paper-roll electrophoresis technique would also require further development in the following procedures, Streaking simple on unrolled, lengthy paper strip, moistening lengthy paper strip with buffer before rolling, rolling up paper strip with alternating special conductive carbon paper terminal at upper and lower edges of paper-roll as well as developing portable polystyrene reservoir for horizontal or vertical orientation of buffer reservoir. Designing connection electrodes for linking DC power source to paper roll, preparation of sandwich paper-roll containing gel and identification of bands of components after separation are also necessary. Machines needed for cutting off segments of paper-roll and extraction of isolated amino acids or proteins from segments of paper-roll have to be made in order to simplify the various tasks required in this large scale technique.

Summary: This new techniques the use of paper roll for either chromatography or electrophoresis with the use of suitable adsorbent paper such as filter paper, paper towel, tea filter or Whatman paper in large-scale separation of streak of organic mixtures using ascending or descending order and suitable polar or non-polar solvents. Separated components appear as bands on the developed chromatogram paper which can be cut into segments according to the number of components on the paper. Invisible components can be identified under ultraviolet light and their locations are marked before cutting the chromatogram roll into segments. Conventional paper and thin layer chromatography may be done prior to paper-roll technique to determine the Rf values of components. However, components can also be located nondestructively by laying one side of the wet, cylindrical chromatogram on a dry thin strip of filter paper while applying little pressure to allow bands of separated components to stain the paper. The paper strip is separated and sprayed with locating agents. Corresponding bands on sprayed paper are aligned with those on the dry paper roll to locate components before the roll is cut into segments. Each component is eluted from the cut out segments with any suitable solvent. Paper roll may be wrapped alone or as a

17

sandwich of paper with silica gel, alumina or calcium carbonate. Buffer solutions and a supply of direct current is required if this technique is used for the electrophoresis separation of ionic organic mixtures containing amino acids and proteins.

Glossary

Terms used in Paper-roll Chromatography/Electrophoresis:

Chromatography is the separation of organic mixtures by using the partition ability of the components between mobile and stationary phases.

Stationary Phase is a substance that is maintained in a fixed position during chromatography.

Mobile Phase is a substance which moves within another stationary substance.

Analyte is the organic mixture that is separated during chromatography.

Solvent is any substance that dissolves a component.

Solute is the particulate mixture which requires separation.

Band is the line of a separated component that appears distinctly on the chromatogram after chromatography.

Chromatogram is bands of separated components which appear on the adsorbent substance after chromatography.

Eluent is a solvent that is used to dissolve components from the cut out segments of the paper roll after chromatography.

Eluate is the liquid mobile phase which leaves the paper roll during descending paper-roll chromatography process.

Component is a substance which has a specific retardation (or retention) factor.

Paper-roll is formed by wrapping a lengthy sheet of paper around a thin cardboard or polymer core.

Paper-roll sandwich is formed by wrapping one or two lengthy sheet(s) of paper containing silica gel, electrophoresis gel, alumina or calcium

carbonate with the rolls which fold around a thin cardboard or polymer core.

Preparative Chromatography is a process used to produce sufficient quantity of separated components.

Analytical Chromatography is a process used to assess the presence and quantity of substances in the separated mixture.

Retardation factor, Rf is the ratio of distance moved by the center of a component to the distance moved by the solvent front.

Locating agent is a substance that is used to determine the position of components on the chromatogram.

Detector is the instrument that is used to identify and measure concentration of components after separation.

T-handle is a T-shaped polymer or glass rod which is used to hang the paper roll within the center of the chromatography chamber.

Chromatography chamber is the container which holds the mobile and stationary phases during chromatography.

References:

1. JOURNAL OF RESEARCH of the National Bureau of Standards- A. Physics and Chemistry Vol. 71 A, No. 1, January-February 1967 Large-Scale, Preparative Paper Chromatography Harriet L Frush Institute for Materials Research, National Bureau of Standards, Washington D.C. 20234 (September 8, 1966)

2. Instructions for LKB "Chromax" Pressurized Paper Chromatography Column System, LKB-Produkter, Stockholm 12, Sweden.

3. H. H. Brownell, J. G. Hamilton, and A. A. Casselman, Anal. J. N. Balston and B. E. Talbot, A Guide to Filter Paper and Cellulose Powder Chromatography (H. Reeve Angel & Co., Ltd., London and W. & R. Balston Ltd., Maidstone, England).

4.www.wikipedia.org/wiki/paper chromatography

5. Mikhail Tswett (1906) "Physikalish-ChemischeStudienuber das Chlorophyll Die Adsorption" Berichte der DeustchenbotanischenGesellschft, vol. 24, pp. 316-326, Tsvet Westheimer, Reiner.

Electrophoresis in Practice. 3rd Ed. New York: Springer Verlag2001.

6. Chris Conoley and Phil Hills, Chemistry (Haper Collins Publishers Ltd., London 2002).

About the author

Akintunde M. Lawal is a freelance scientist, with a postgraduate degree in Pharmacy. He is the author of a Sci-Fi novel titled 27th Century Fiasco.

Contact: akintundemlawal@gmail.com

Watch out for series of demonstration projects caption 'Paper-roll Chromatography & Electrophoresis' on YouTube very soon.